# Counting at the z[oo]

- **Count up to 20 objects**

3

## Teacher's notes

Look at the picture. Count how many there are of each animal and write the number in the correct box.

3

Date_____

# Apple addition

● **Add two numbers**

| | | |
|---|---|---|
| 4 + 1 = | 2 + 1 = | 1 + 4 = |
| 1 + 3 = | 1 + 2 = | 3 + 1 = |

2 + 1 = 3   is the same as

is the same as

is the same as

## Teacher's notes

4

Match each of the addition sums at the top of the page with the correct picture.
Write the sum and the answer in the boxes provided.

Date _____

# Jumping frogs

- **Add two numbers**

## Jump on **1** jump

| 1 | + | 1 | = | 2 |

 1 2 3 4 5 6 7 8 9 10

## Jump on **2** jumps

| 1 | + | | = | |

 1 2 3 4 5 6 7 8 9 10

## Jump on **3** jumps

| 1 | + | | = | |

1 2 3 4 5 6 7 8 9 10

## Jump on **4** jumps

| 1 | + | | = | |

 1 2 3 4 5 6 7 8 9 10

## Jump on **6** jumps

| 1 | + | | = | |

 1 2 3 4 5 6 7 8 9 10

## Teacher's notes
Count how many lily pads the frog will jump and then write the sum in the spaces provided.

Date_____

# Adding in the garden

- Add two numbers
- Write addition number sentences

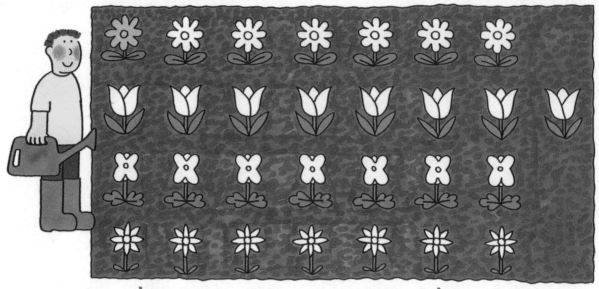

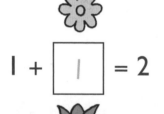

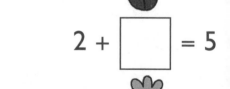

1 + ☐1☐ = 2

0 + ☐ = 3

2 + ☐ = 5

1 + ☐ = 3

2 + ☐ = 4

4 + ☐ = 5

2 + ☐ = 3

3 + ☐ = 4

3 + ☐ = 5

### Teacher's notes

6

Write the missing number in each of the addition calculations. Match the flower above it to those in the garde
The missing number tells you how many of each flower to colour in.

Date_____

# Cake subtraction

● **Understand subtraction as 'taking away'**

$\boxed{4}$ − 3 = ◯

$\boxed{\phantom{0}}$ − 3 = ◯

$\boxed{\phantom{0}}$ − 2 = ◯

$\boxed{\phantom{0}}$ − 1 = ◯

$\boxed{\phantom{0}}$ − 2 = ◯

$\boxed{\phantom{0}}$ − 1 = ◯

$\boxed{\phantom{0}}$ − 2 = ◯

$\boxed{\phantom{0}}$ − 4 = ◯

## Teacher's notes

Count the cakes in each tin and write the number in the square. Cross out the number of cakes you are taking away. Then write in the circle how many cakes are left.

Date_____

# Dice difference

- **Use objects to take away a smaller number from a larger number**
- **Write subtraction number sentences**

**You need:**
- 2 1–6 dice

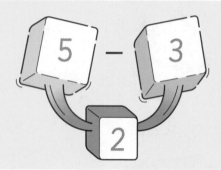

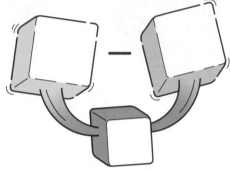

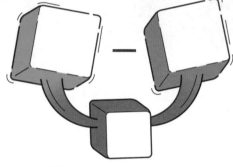

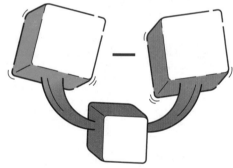

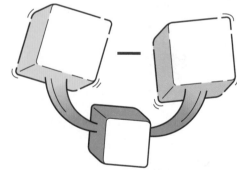

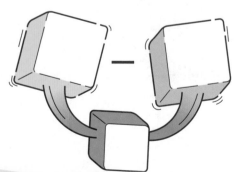

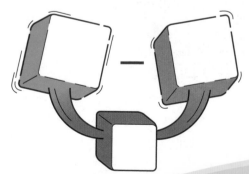

## Teacher's notes

**8**

Roll two dice. Record the two numbers on the larger cubes (put the bigger number first). Write the answer on the smaller cube below.

Date_____

# Coin totals

- Solve problems about money
- Count up to 20 objects

**You need:**
- coloured pencils

is the same as

P

is the same as

P

is the same as

P

is the same as

P

is the same as

P

## Teacher's notes

Look at the coins in each purse. Colour the pennies next to each purse to show the same amount of money. Then write the total for that purse in the box.

Date _____

# Change from 5p

### Solve problems about money

Rob has 5p.  or

He can buy one toy car.

Work out his change for each one.

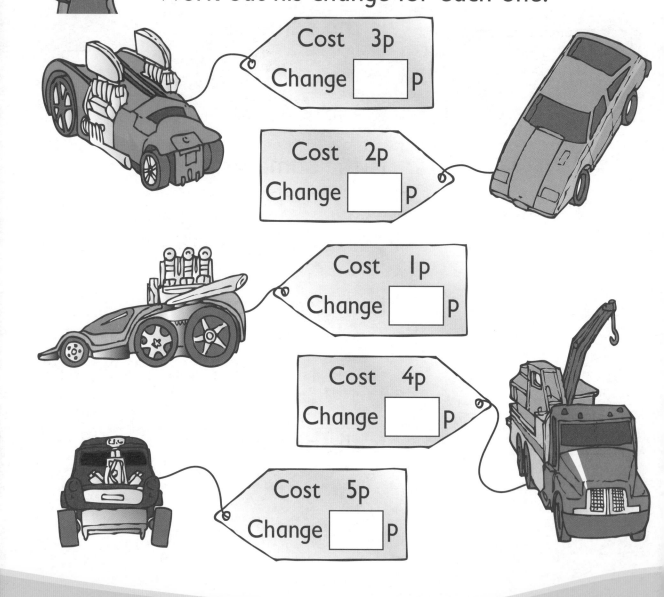

Cost    3p

Change [    ] p

Cost    2p

Change [    ] p

Cost    1p

Change [    ] p

Cost    4p

Change [    ] p

Cost    5p

Change [    ] p

**Teacher's notes**

10

Look at the price of each toy car. How much change would you get from 5p?
Write the answer in the box.

Date _____

# At the shops

**You need:**
- coloured pencils

● **Solve problems about money**

 Ben buys a

5p – $\boxed{3p}$ = $\boxed{2}$ p

 Hanna buys a

5p – $\boxed{\phantom{00}}$ = $\boxed{\phantom{00}}$ p

 Ellie buys a

5p – $\boxed{\phantom{00}}$ = $\boxed{\phantom{00}}$ p

 Fahd buys a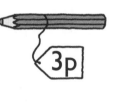

6p – $\boxed{\phantom{00}}$ = $\boxed{\phantom{00}}$ p

 Mia buys a

6p – $\boxed{\phantom{00}}$ = $\boxed{\phantom{00}}$ p

## Teacher's notes
Each child has either 5p or 6p to spend. First, look at what the children bought and work out the change.
Then, in the purse, colour the coins they might have received as change.

11

Date_____

# Guess how many

- **Estimate a set of objects**

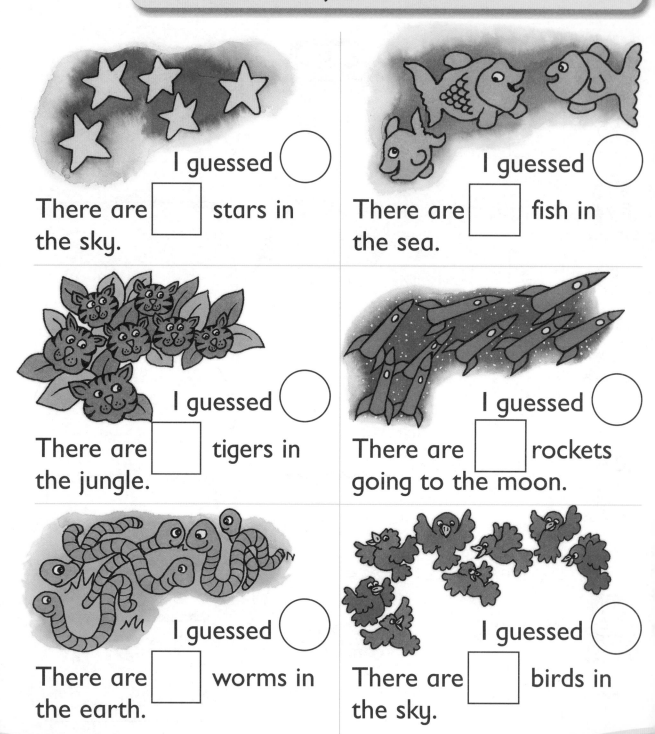

I guessed ◯

There are ☐ stars in the sky.

I guessed ◯

There are ☐ fish in the sea.

I guessed ◯

There are ☐ tigers in the jungle.

I guessed ◯

There are ☐ rockets going to the moon.

I guessed ◯

There are ☐ worms in the earth.

I guessed ◯

There are ☐ birds in the sky.

## Teacher's notes

**12**

Quickly look at each set of objects. Cover the objects with your hand and write your estimate in the circle. Now count the number of objects. Write your answer in the box.

# Jumping frogs addition

Add numbers with answers up to 5

**Teacher's notes**
Draw a line to match the calculation on each frog with the correct answer on a lily pad.

13

Date _____

# Aladdin's addition

● **Add numbers with answers up to 5**

0 + [2] = 2

1 + [ ] = 2

2 + [ ] = 2

0 + [ ] = 3

1 + [ ] = 3

2 + [ ] = 3

3 + [ ] = 3

0 + [ ] = 4

1 + [ ] = 4

2 + [ ] = 4

3 + [ ] = 4

4 + [ ] = 4

0 + [ ] = 5

1 + [ ] = 5

2 + [ ] = 5

3 + [ ] = 5

4 + [ ] = 5

5 + [ ] = 5

**Teacher's notes**

14

In each calculation, write in the missing number.

Date_____

# Cake calculations

● **Add and subtract numbers**

**You need:**
● red coloured pencil

$$2 + \boxed{1} = 3$$

$$3 - 2 = \boxed{\phantom{0}}$$

$$3 + \boxed{\phantom{0}} = 4$$

$$4 - 3 = \boxed{\phantom{0}}$$

$$2 + \boxed{\phantom{0}} = 4$$

$$4 - 2 = \boxed{\phantom{0}}$$

$$2 + \boxed{\phantom{0}} = 5$$

$$5 - 2 = \boxed{\phantom{0}}$$

$$1 + \boxed{\phantom{0}} = 5$$

$$5 - 1 = \boxed{\phantom{0}}$$

$$4 + \boxed{\phantom{0}} = 6$$

$$6 - 4 = \boxed{\phantom{0}}$$

**Teacher's notes**

In each picture find out how many cakes are to be decorated by drawing in the missing cherries.
Then complete the addition and subtraction calculations. You can use the row of cakes like a number line.

15

Date_____

# Juggle the jugs

• **Subtract numbers**

**Teacher's notes**

Write in the − and = signs in the correct places. Now look at each row of jugs. Unscramble the calculations on the left-hand side and write them in the correct order on the right-hand side.

Date _____

# Shipshape ships

- **Recognise 2-D shapes**

## Teacher's notes

Look at each ship and count how many coloured squares, triangles, circles and rectangles there are.
Write the total number of each shape in the correct shape underneath.

17

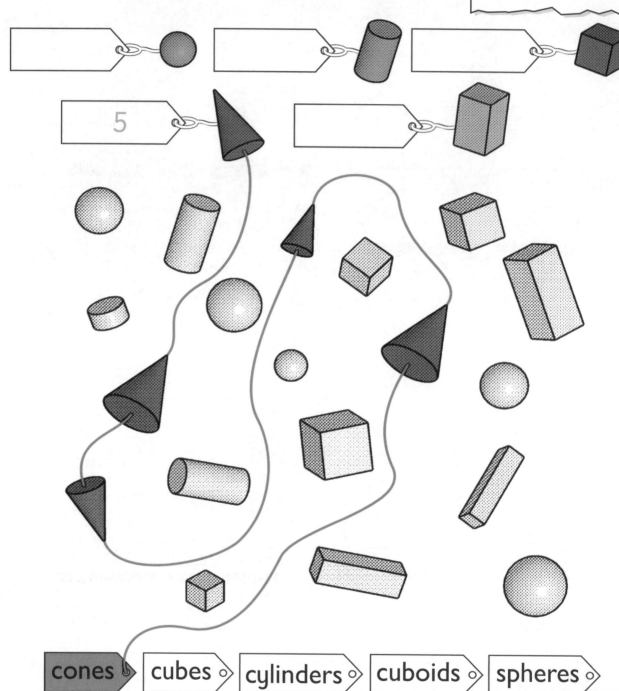

Date _____

# Match the beads

**Recognise and name 3-D solids**

**You need:**
- coloured pencils

5

cones ▷ cubes ◇ cylinders ◇ cuboids ◇ spheres ◇

## Teacher's notes

Colour the beads in their shape colour. Draw a line to join the beads of the same shape and write the total number of shapes on the shape's label. Continue the line to the name label and colour it to match.

Date _____

# Sort the shapes!

- Name 2-D and 3-D shapes
- Use 2-D and 3-D shapes to make patterns, pictures and models

The Shape Shop needs to be sorted out! Can you put all the shapes back onto the right shelves?

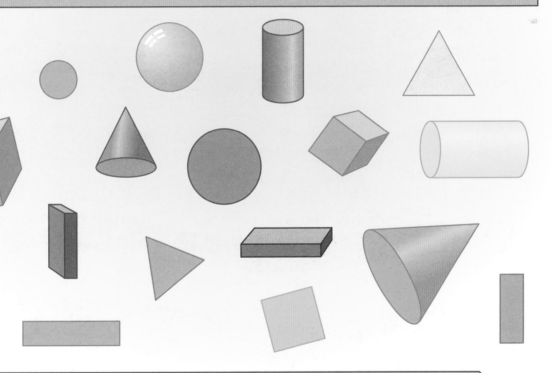

### Teacher's notes

Look at the shapes. Draw a line to sort each flat shape onto the top shelf and each solid shape onto the bottom shelf.

19

Date_____

# Sari patterns

● Use shapes to make a
simple pattern

**You need:**

● coloured pencils

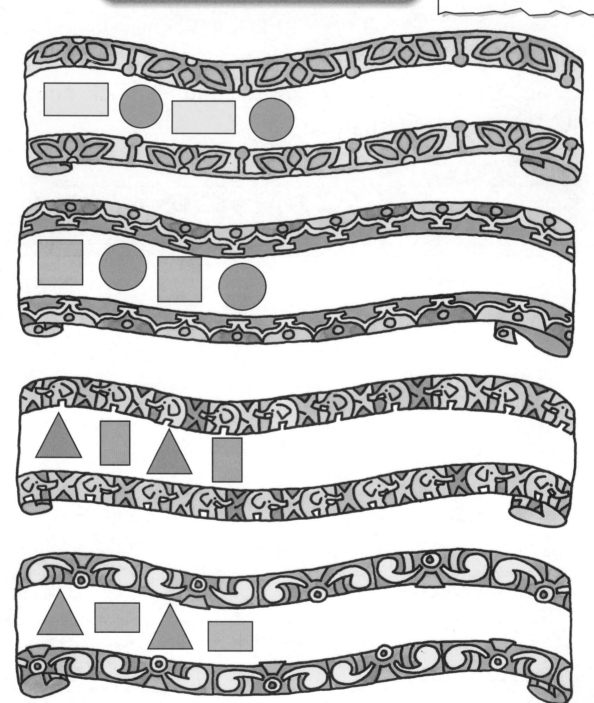

## Teacher's notes

**20**

Look at the pattern on each sari. Now continue the pattern by drawing and colouring the next four
shapes on each sari to match the first four shapes.

Date _____

# Jungle counting

- Count at least 20 objects

| 8 |  lizards | |  snakes | |  birds |

| |  bugs | |  butterflies | |  tigers |

| |  monkeys |

## Teacher's notes

Count how many there are of each creature and write the answer in the correct box.

Date_____

# One more, one less

Find the number that is one more or one less than a number

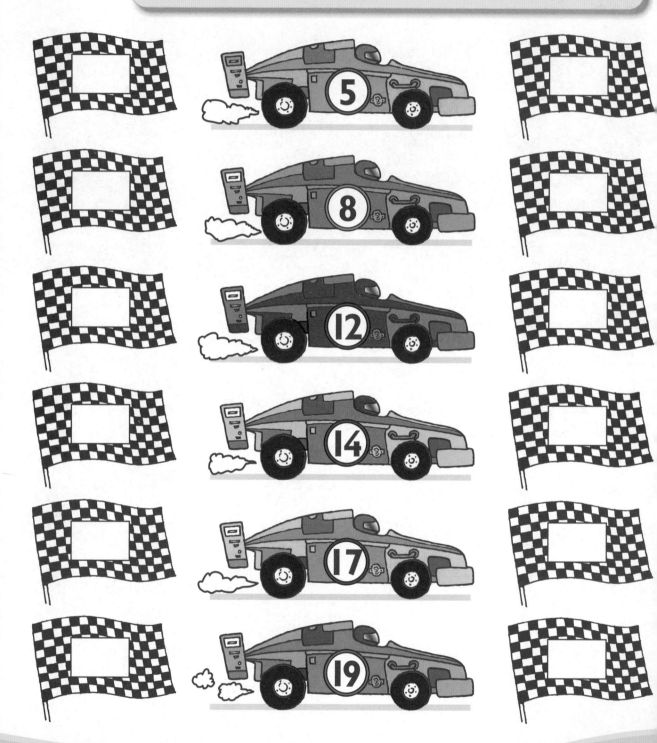

## Teacher's notes

Look at the number on each car. On the left-hand side write the number that is one less than that number, and on the right, the number that is one more.

22

Date_____

# Flower counting

Count on and back in ones

**Teacher's notes**
In each row, count on or back in ones and write the missing numbers in the boxes.

23

Date_____

# Counting heads

● **Know pairs of numbers with a total of 10**

## Teacher's notes

Ten children live on each storey of two blocks of flats. Draw faces at each of the two windows to total 10.
Find as many ways as you can.

Date _____

# Flower petal addition

● **Solve a puzzle**

You need:
● coloured pencils

## Teacher's notes

Colour two petals on each flower that together add up to the total in the middle. If there is more than one answer, use different colours for each calculation.

Date _____

# Space problems

Solve a problem

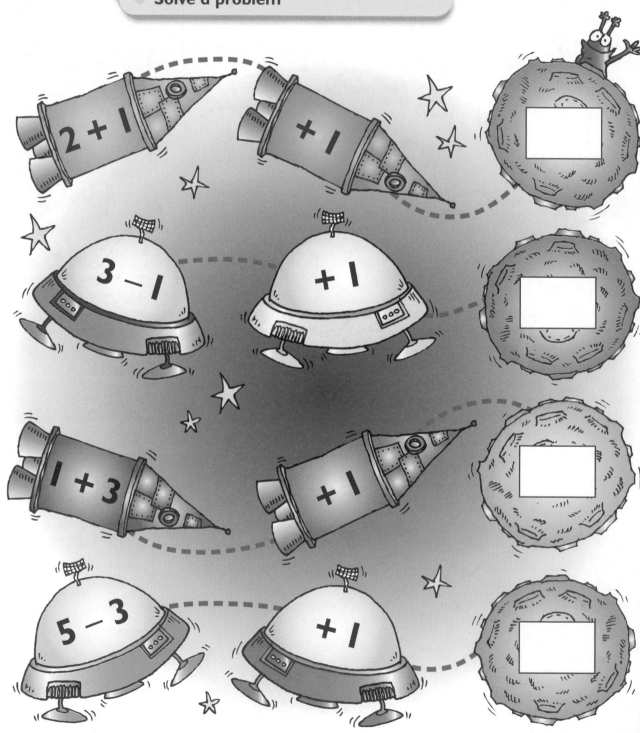

## Teacher's notes

Follow the instructions on the spaceships to find out how many aliens land on each planet.

Date _____

# Shortest to longest

● Compare the lengths of more than two objects and put them in order

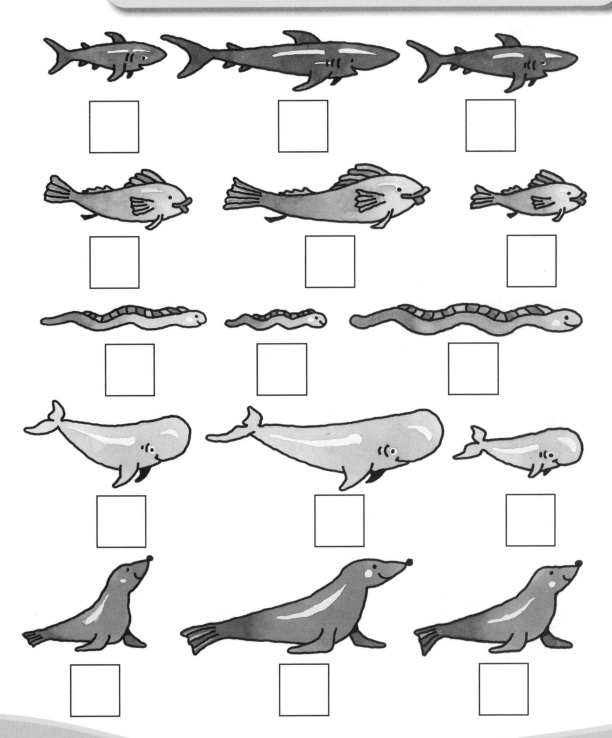

## Teacher's notes

Label each row of sea creatures from the shortest to the longest: 1, 2 or 3.
(Label the shortest 1 and the longest 3.) Then circle the longest.

Date_____

# Measure in spans

### Estimate, measure and compare objects

This is my hand span. Now measure yours on top!

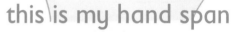

this is my hand span

table

I measured ☐ spans.

book

I measured ☐ spans.

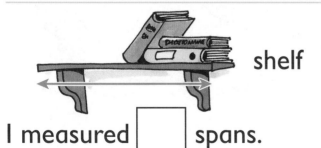

shelf

I measured ☐ spans.

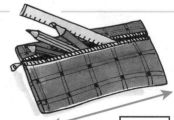

pencil case

I measured ☐ spans.

### Teacher's notes

**28**

Measure and record your hand span using the character's span as a guide and comparison.
Then measure different objects around the classroom, recording the number of spans each time.

Date_____

# More or less

Measure and compare objects

You need:

● metre rule

| Name | More than a metre | Less than a metre |
|------|-------------------|-------------------|
|      |                   |                   |
|      |                   |                   |
|      |                   |                   |
|      |                   |                   |
|      |                   |                   |
|      |                   |                   |
|      |                   |                   |

How many are more than a metre? ☐

How many are less than a metre? ☐

**Teacher's notes**

Work in groups. Put your names in the table. Now measure each other with a metre rule.
Tick the second column if you are more than the metre rule. Tick the third column if you are less.

Date_____

# Estimate and measure

**Measure and compare objects**

You need:
- cubes

| Object | Estimate | Measurement |
|---|---|---|
| | 5 cubes | 3 cubes |
| | | |
| | | |
| | | |
| | | |
| | | |
| | | |

## Teacher's notes

30

Choose a uniform non-standard measure, such as a set of cubes. Estimate the length of each object as so many cubes. Write the number of cubes and check the estimate.

Date_____

# Snake lengths

**Measure and compare objects**

You need:
● counters

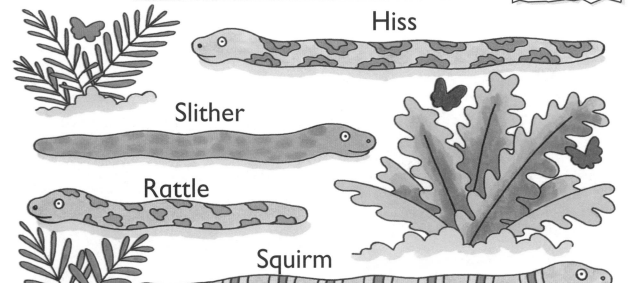

Hiss

Slither

Rattle

Squirm

How long is Hiss?

[  ] counters long

How long is Squirm?

[  ] counters long

How long is Slither?

[  ] counters long

How long is Rattle?

[  ] counters long

How much longer is Hiss than Rattle? [  ] counters

Write down the snakes in order, longest to shortest.

_____ _____ _____ _____

**Teacher's notes**

Measure each of the snakes using counters and answer the questions.

Date _____

# Toy shop sorting

Sort objects

**You need:**
- small and large coloured counters
- coloured pencils

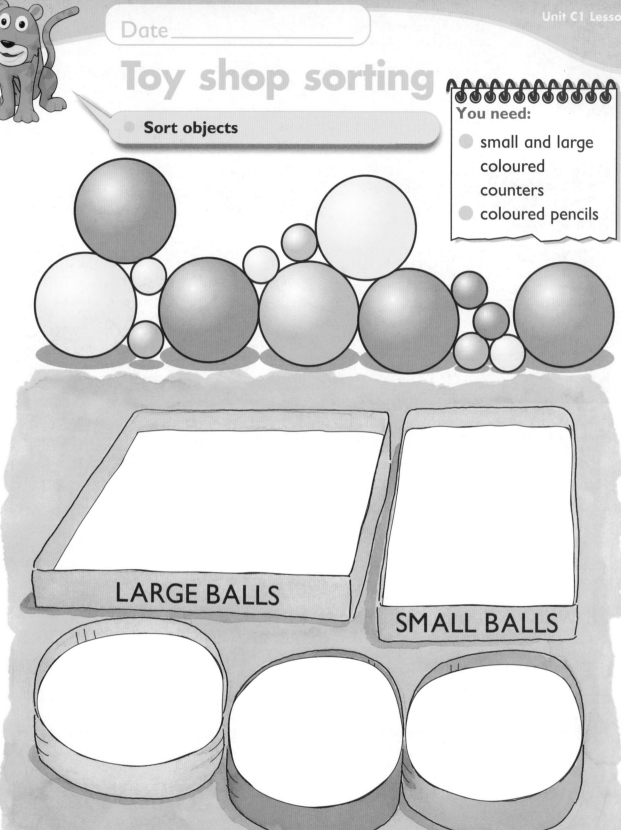

LARGE BALLS

SMALL BALLS

## Teacher's notes

**32**

Place matching counters on the balls one at a time.
Sort the counters by size by putting them in the boxes marked 'LARGE BALLS' and 'SMALL BALLS'.
Draw round the counters and colour them. Repeat, sorting the counters by colour into the round boxes.

# Sorting at the market stall

Sort objects and make a table

pears

| pears | 7 |
| apples | |
| oranges | |
| bananas | |

Which fruit is there most of?

## Teacher's notes

Look at the fruit and draw the same number of fruit in the boxes. Label the boxes.
Count the fruit and complete the table. Then answer the question.

33

Date_____

# Chocolate shapes

- Use diagrams to sort objects

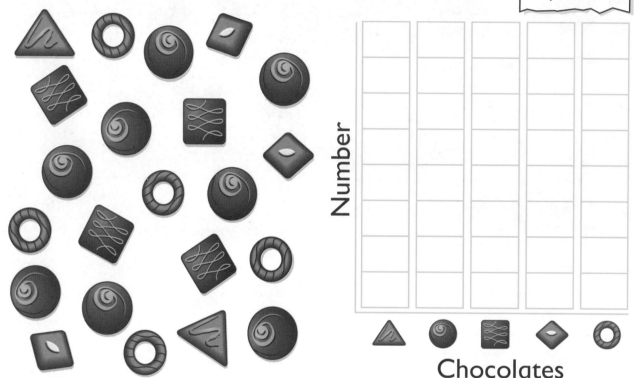

Number

Chocolates

There are _____ square chocolates.

There are 8 _____ chocolates.

The least common shape is _____.

There are more _____ than _____ chocolates.

| Shape | ▲ | ● | ▣ | ◆ | ◯ |
|-------|---|---|---|---|---|
| Number | | | | | |

**Teacher's notes**

34

For each chocolate, cross it off and colour a square in the block graph. Answer the questions.
Complete the table.

Date_____

# Lists of animals

- **Sort objects using a list**

| sheep | bees | crabs | cow | horses |
|---|---|---|---|---|

| butterflies | fish | dolphin | earwigs |
|---|---|---|---|

| Farm animals | Insects | Sea creatures |
|---|---|---|
| 2 horses | 2 earwigs | 1 dolphin |
| _____ | _____ | _____ |
| _____ | _____ | _____ |

There are ☐ farm animals.   There are ☐ insects.

There are ☐ fish.   There are ☐ animals altogether.

**Teacher's notes**

Look at the animals. Make a list of the **farm animals**, the **insects** and the **sea creatures** using the names at the top to help you. Then complete the sentences underneath.

35

Date_____

# Families

Use diagrams, lists and tables to sort objects into groups

**You need:**
- coloured pencils

Pet

No pet

## Sisters in the family

No sister

One sister

Two sisters

_____ families have no pet.

There are _____ children with no sisters.

**Teacher's notes**

Look at the 8 families. Look at each block and colour one square for each family it applies to.

Date _____

# Before or after?

● **Know the days of the week and say them in order**

| comes before | | comes after |

---

Monday

comes before

Tuesday

---

Wednesday

Tuesday

---

Wednesday

Thursday

---

Thursday

Friday

---

Saturday

Friday

---

Sunday

Saturday

---

## Teacher's notes

Look at what Harry is planning to do on each day of his holiday. Write 'comes before' or 'comes after' in the boxes between the pictures.

Date_____

# Months of the year

**Order the months of the year**

January

April | June

August | September

October

My birthday is in the month of [              ].

## Teacher's notes

**38**

Write the missing months of the year in the spaces on the train carriages and on the birthday cakes.
Complete the sentence below.

Date _____

# Round the year

Know the seasons of the year and say them in order

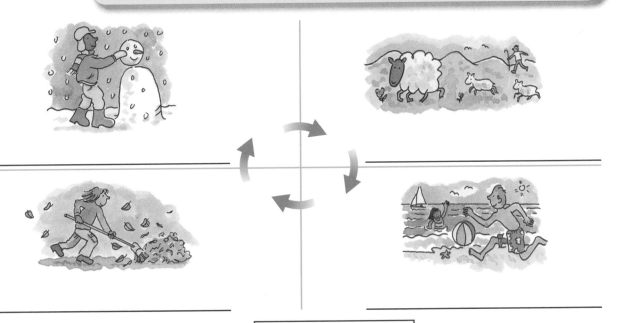

Winter comes | before | Spring

Summer comes | after | Spring

Spring comes | | Winter

Winter comes | | Autumn

Autumn comes | | Winter

Spring comes | | Summer

Summer comes | | Autumn

**Teacher's notes**

*Top:* Write the season underneath each picture.
*Bottom:* Write 'before' or 'after' in the boxes.

Date_____

# Tell me the time!

● **Read the time to the hour**

| I o'clock | 7 o'clock | II o'clock |

## Teacher's notes

**40**

*Top:* Write the time underneath the clock faces.

*Middle:* Draw the times on the clock faces.

*Bottom:* Draw and write times on the hour. Choose times that are different from those above.

Date_____

# My diary

• **Know the days of the week and read time to the hour**

| **Monday** | |
|---|---|
| Assembly 9 o'clock | Football 6 o'clock |
| | **Saturday** |
| Swimming 5 o'clock | Ahmed's birthday |
| | party 1 o'clock |
| Emma's 4 o'clock | |
| **Thursday** | Granny's 3 o'clock |
| | |

Assembly   Football

Swimming   Ahmed's birthday party

Emma's   Granny's

## Teacher's notes

*Top:* Write the missing days in the diary. Decide what to do on Thursday and write it underneath.
*Bottom:* Fill in the hands on the clock faces for the events shown in the diary.

41

Date_____

# Supermarket estimation

● **Estimate a group of objects**

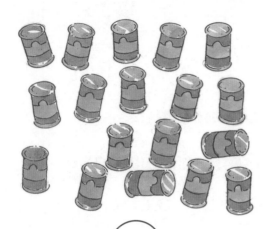

I estimate ◯

There are ☐ tins.

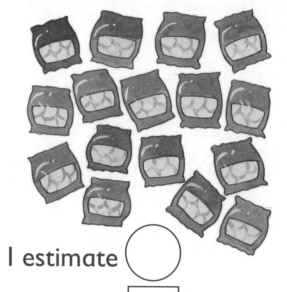

I estimate ◯

There are ☐ packets.

I estimate ◯

There are ☐ tomatoes.

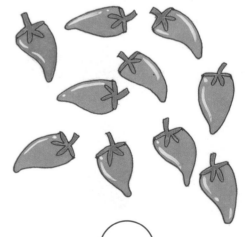

I estimate ◯

There are ☐ okra.

## Teacher's notes

Quickly look at each set of objects. Cover the objects with your hand and write your estimate in the circle. Now count the number of objects. Write your answer in the box.

Date _____

# Centipede centimetres

● Measure objects

You need:
● ruler

 cm

cm

 cm

 cm

 cm

 cm

## Teacher's notes

Look at each centipede in turn and use a centimetre ruler to measure the length of each.
Write the answer in the box.

43

Date _____

# Which picture?

● **Describe the position of an object**

**You need:**
● coloured pencils

## What is ...

on the right of the
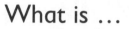

on the left of the

between the  and the

between the

under the

above the

and the

**Teacher's notes**

**44**   Look at the pictures on the picture board to answer the questions. Draw in the picture which gives the answer.

Date _____

# The way home

- ● **Decribe position and direction**

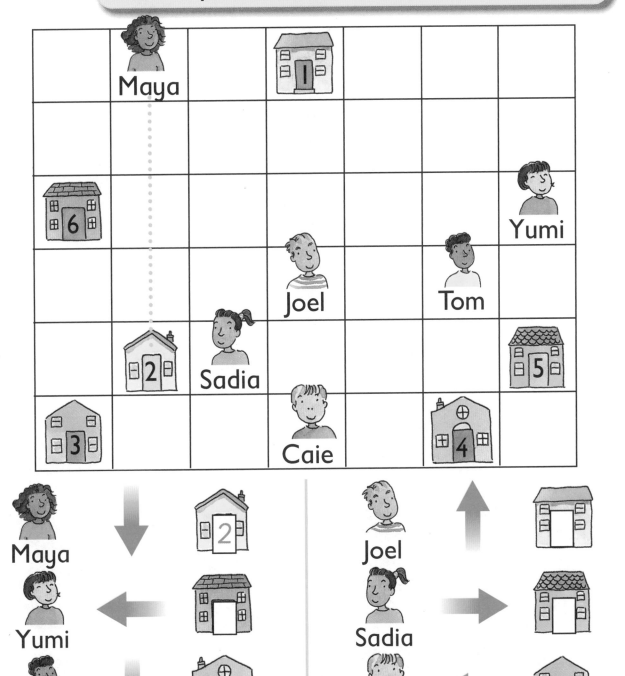

## Teacher's notes

*Top:* Use the clues at the bottom of the page to help you decide where each child lives. Draw a line to join each child to their house. *Bottom:* Write their house number on the house next to their name.

45

Date_____

# Turning around

● **Recognise and make whole and half turns**

## After a whole turn

Tom sees

Yee sees

Kabir sees

Sarai sees

## After a half turn

Tom sees

Yee sees

Kabir sees

Sarai sees

## Teacher's notes

**46**

Look at the picture at the top. Below, draw a line from each child to what they would see after a whole turn of the roundabout. Then repeat, drawing a line to what they would see after half a turn.

Date_____

# Train 2s

● Count on and back in twos

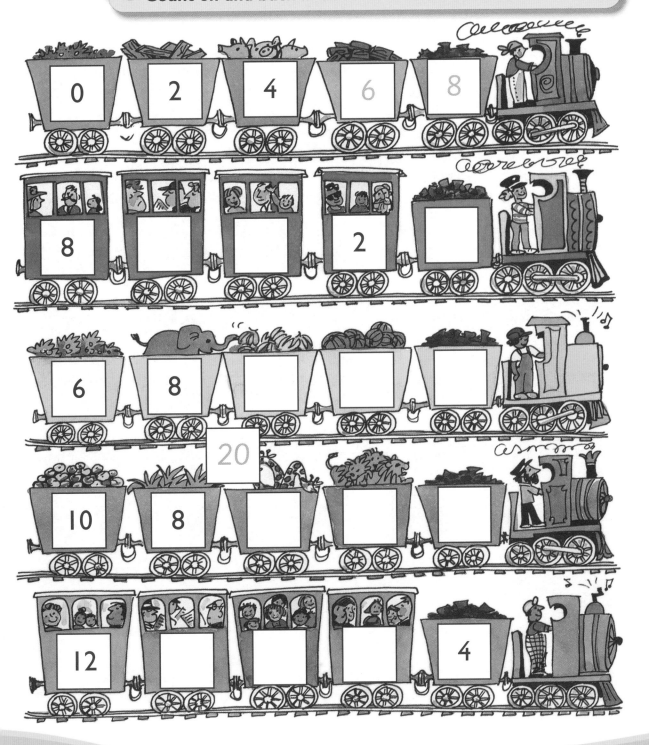

**Teacher's notes**

In each row, count on or back in 2s and write the missing numbers in the boxes on the carriages.

Date_____

# Flying fives

**Count on in fives**

**You need:**
- red coloured pencil

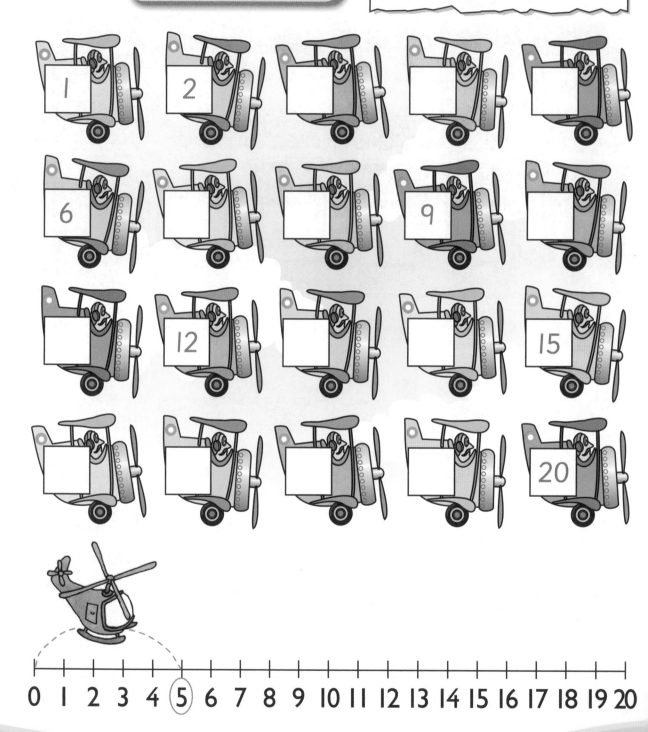

0 1 2 3 4 ⑤ 6 7 8 9 10 11 12 13 14 15 16 17 18 19 20

## Teacher's notes

48

Complete the number track. Then count on from zero in steps of five, colouring each multiple of five in red.
Underneath make the helicopter land on 20 in steps of five, circling the numbers that it lands on.

Date _____

# Steps of 2 and 5

● **Count on in twos and fives**

| 10 | 12 | | | |

| 0 | | 10 | | |

| 20 | 22 | | | |

| 10 | | | | 30 |

## Teacher's notes

Complete the number sequences by counting on in steps of two or five.

49

Date_____

# Terrible tens!

● Count on in tens

## Teacher's notes

**50**

Look at the number in each each monster's mouth and count on or back in tens, writing the missing numbers in the empty mouths.

Date _____

# Number patterns

Count on in twos, fives and tens

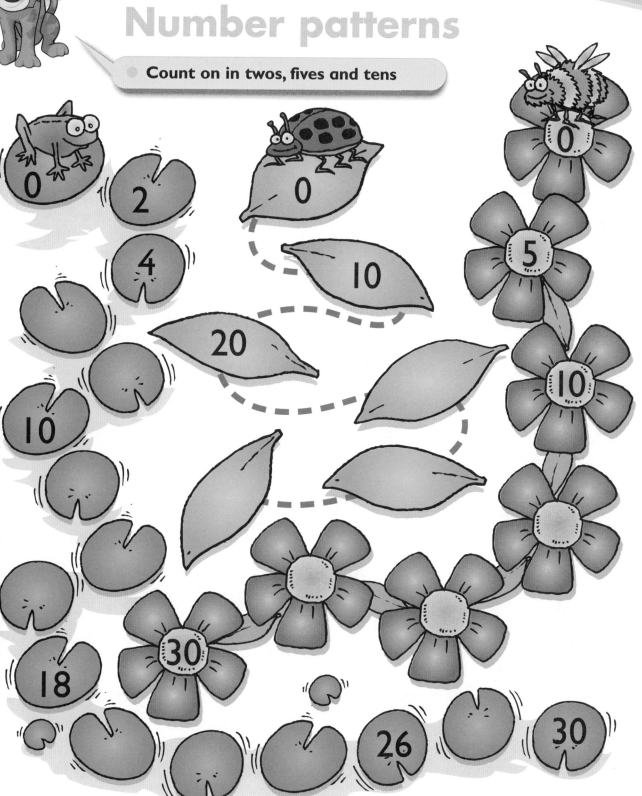

**Teacher's notes**

In each chain, count on in twos, fives or tens and write the missing numbers in the spaces provided.

51

Date _____

# Pizza portions

Find half of a shape

You need:
- coloured pencils

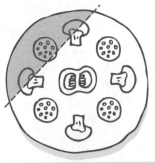 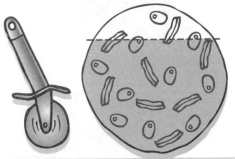 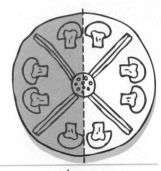

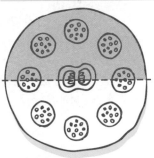 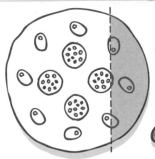  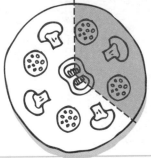

 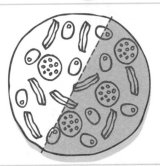 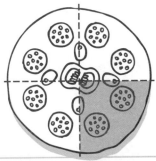 

  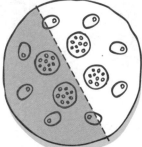

## Teacher's notes

In each row, colour the pizza that has been cut exactly in half.

**52**

Date _____

# Half time halves

● **Find half of a set of objects**

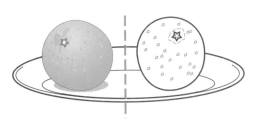

Half of [ 2 ] is [ ]

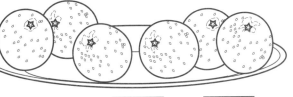

Half of [ 6 ] is [ ]    Half of [ ] is [ ]

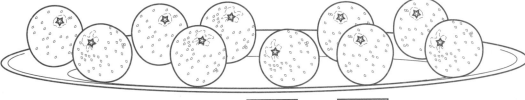

Half of [ ] is [ ]

Half of [ ] is [ ]

## Teacher's notes

The footballers are eating oranges at half time. The oranges need to be shared equally between the two teams. Count the oranges on each plate then draw a line to show two halves on each plate. Colour one half on each plate in orange and complete each sentence below.

53

Date_____

# How many are there?

**Solve simple word problems**

Sam has
2 cats.

Ellis has
2 cats.

How many cats do they
have altogether?

   =

There were 10 socks
on the washing line.

3 blew
away
in the wind.

How many socks were left?

  =

How much money is in
Laura's purse?

   =

Kate has
6 pencils.

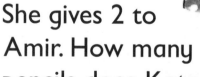

She gives 2 to
Amir. How many
pencils does Kate have left?

   =

## Teacher's notes

**54** Read each problem. Then write the correct addition or subtraction calculation for each problem in the
boxes underneath.

Date_____

# Number problems

● Solve simple word problems

 There were 6 cherries.
Samir ate 4.

How many were left?

□ ○ □ = □

Mo has 8 cards.

Jack has 2.

How many cards do
they have altogether?

□ ○ □ = □

 Pip had 5 bones. He dug up 3 more.

How many bones does Pip have?

□ ○ □ = □

Jenny had
10 marbles.

She lost 5.

How many marbles
were left?

□ ○ □ = □

## Teacher's notes

Read each problem. Then write the correct addition or subtraction calculation for each problem in the
boxes underneath.

Date_____

# Apple addition and subtraction

### Write addition and subtraction number sentences

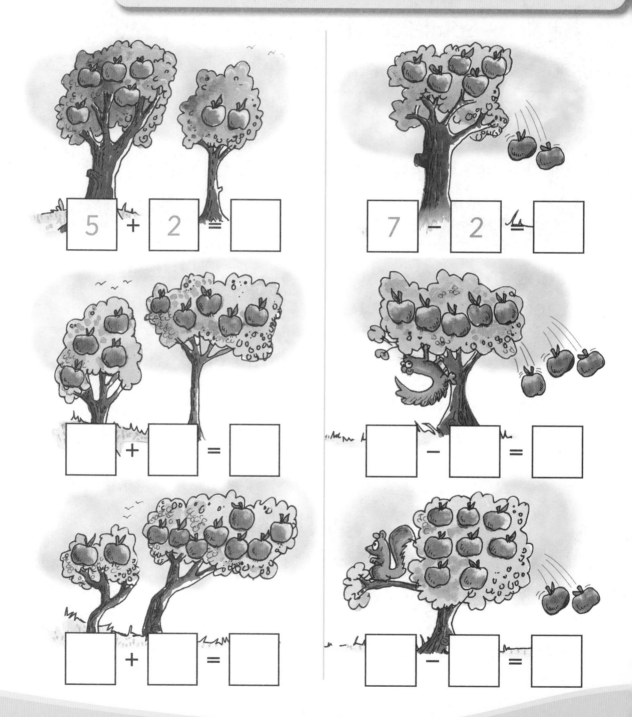

$\boxed{5} + \boxed{2} = \boxed{\phantom{0}}$

$\boxed{7} - \boxed{2} = \boxed{\phantom{0}}$

$\boxed{\phantom{0}} + \boxed{\phantom{0}} = \boxed{\phantom{0}}$

$\boxed{\phantom{0}} - \boxed{\phantom{0}} = \boxed{\phantom{0}}$

$\boxed{\phantom{0}} + \boxed{\phantom{0}} = \boxed{\phantom{0}}$

$\boxed{\phantom{0}} - \boxed{\phantom{0}} = \boxed{\phantom{0}}$

## Teacher's notes

**56**

Look at the apples on the trees in the first column and write the corresponding addition number sentence. Then look at the apples on the trees in the second column. Count the total number of apples, then subtract the number of apples that have fallen off the tree and write the corresponding subtraction number senten

Date _____

# Double bubbles

Know doubles of numbers to 5 + 5

You need:
- coloured pencils

$$1 + 1 = 2 \qquad \Box + \Box = \Box$$

$$\Box + \Box = \Box$$

$$\Box + \Box = \Box \qquad \Box + \Box = \Box$$

## Teacher's notes

Look for the two bubbles which are the same and the bubble with their total. Colour all three bubbles in the matching colour and write the doubles calculation in the matching coloured boxes.

57

Date _____

# Caterpillar calculations

### Understand addition and subtraction

$2$ + $1$ = $3$

$3$ − $1$ = 

---

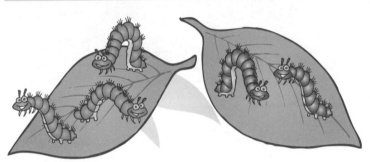

 + = 

 − = 

---

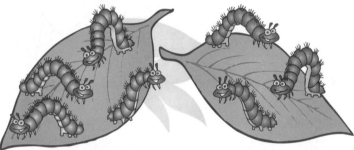

 + = 

 − = 

---

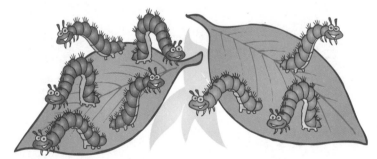

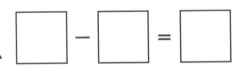

 + = 

 − = 

---

**Teacher's notes**

In each row, children write the addition calculation and then the matching subtraction calculation.

Date _____

# Plant pot sums

● **Work out the missing number in a number sentence**

## Teacher's notes

Solve each addition and subtraction calculation. Write the answer on the plant or on the pot.

Date _____

# Ladybird subtraction

● **Write subtraction number sentences**

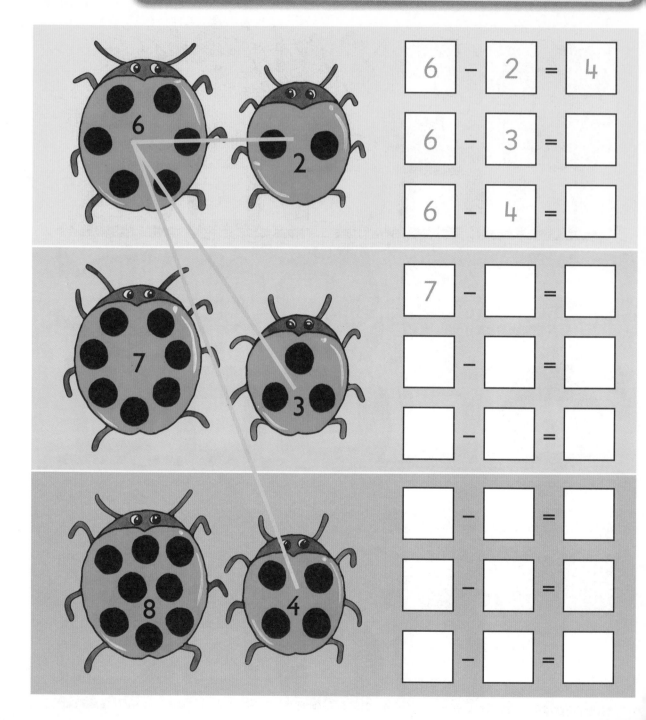

| | | | | |
|---|---|---|---|---|
| 6 | − | 2 | = | 4 |
| 6 | − | 3 | = | |
| 6 | − | 4 | = | |

| | | | | |
|---|---|---|---|---|
| 7 | − | | = | |
| | − | | = | |
| | − | | = | |

| | | | | |
|---|---|---|---|---|
| | − | | = | |
| | − | | = | |
| | − | | = | |

## Teacher's notes

Subtract the number of spots on each small ladybird from the number of spots on each large ladybird. Write the calculation, putting the larger number first.